WILLIAM WOODVILLE, M.D. F.R.S.

Author of Medical Botany, &c.

View of the inoculating Hospital at Pancras

Title: **MEDICAL & OFFICINAL PLANTS - VOL. 3 By William Woodville & James Sowerby XIX centuries engravings**
Series realized by Luca S. Cristini. Scientific consulence by Marco Rampinelli.

ISBN code: 978-88-93272308 First edition March 2017
Code.: **MUSEUM-006**, Editorial series code: Darwin's View **DV-005**

Cover & Art Design: Luca S. Cristini & Anna Cristini
MUSEUM is a trademark of Soldiershop publishing, via Padre Davide, 7 - 24050 Zanica (BG) ITALY. www.bookmuseum.it

WILLIAM WOODVILLE - JAMES SOWERBY

# MEDICAL & OFFICINAL PLANTS - VOL. 3

## PIANTE OFFICINALI, MEDICINALI E AROMATICHE

Shown in this series of three books are the complete and original pattern hand-colored engravings plates from the artist James Sowerby's medical plants present in the great work of William Woodville: Medical botany (London : Printed and sold for the author, by James Phillips, 1790-1793). Medical Botany, William Woodville's three volume work of materia medica, was published in monthly installments between 1790 and 1793.

A third edition of five volumes (the same used in our reproduction) was presented in 1832, twenty-seven years after Woodville's death. This publication added descriptions of thirty-nine new plants and was edited and revised by the eminent botanist, William Jackson Hooker (1785-1865).

With this work, Woodville intended to educate medical practitioners about the plants they prescribe and improve upon preceding works by introducing new plants and more detail.

William Woodville was an English physician and botanist who lived and worked in London for most of his life. For his work in botany, he was made a Fellow of the Linnaean Society just a year after the publication of volume one of his Medical Botany.

DARWIN'S
VIEW

MUS
EUM
BIBLIOTHECA

# MEDICAL BOTANY

Containing systematic and general descriptions, with plates, of all the medicinal plants, indigenous and exotic, comprehended in the catalogues of the *materia medica*, as published by the Royal colleges of physicians of London and Edinburgh: accompanied with a circumstantial detail of their medicinal effects, and of the diseases in which they have been most successfully employed.

By William Woodville, M. D. Of the Royal college of physicians, London. Printed and Sold for the Author, by James Phillips, in five volumes (Edition of 1832)

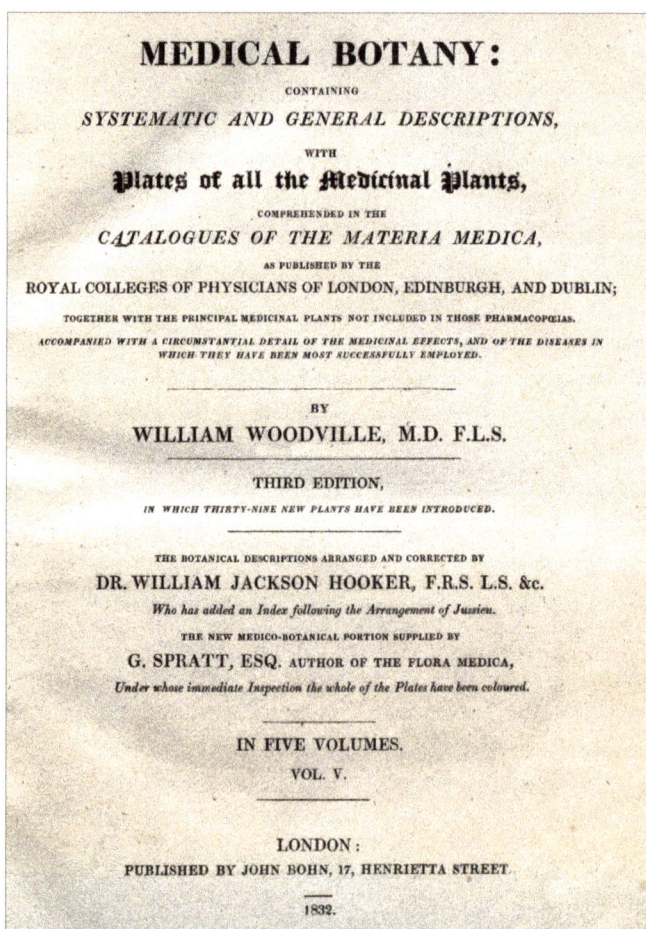

**MEDICAL BOTANY:**

CONTAINING

*SYSTEMATIC AND GENERAL DESCRIPTIONS,*

WITH

**Plates of all the Medicinal Plants,**

COMPREHENDED IN THE

*CATALOGUES OF THE MATERIA MEDICA,*

AS PUBLISHED BY THE

ROYAL COLLEGES OF PHYSICIANS OF LONDON, EDINBURGH, AND DUBLIN;

TOGETHER WITH THE PRINCIPAL MEDICINAL PLANTS NOT INCLUDED IN THOSE PHARMACOPŒIAS.

ACCOMPANIED WITH A CIRCUMSTANTIAL DETAIL OF THE MEDICINAL EFFECTS, AND OF THE DISEASES IN WHICH THEY HAVE BEEN MOST SUCCESSFULLY EMPLOYED.

BY

**WILLIAM WOODVILLE, M.D. F.L.S.**

THIRD EDITION,

*IN WHICH THIRTY-NINE NEW PLANTS HAVE BEEN INTRODUCED.*

THE BOTANICAL DESCRIPTIONS ARRANGED AND CORRECTED BY

**DR. WILLIAM JACKSON HOOKER, F.R.S. L.S. &c.**

*Who has added an Index following the Arrangement of Jussieu.*

THE NEW MEDICO-BOTANICAL PORTION SUPPLIED BY

**G. SPRATT, ESQ.** AUTHOR OF THE FLORA MEDICA,

*Under whose immediate Inspection the whole of the Plates have been coloured.*

IN FIVE VOLUMES.

VOL. V.

LONDON:

PUBLISHED BY JOHN BOHN, 17, HENRIETTA STREET.

1832.

*Medicus omnium Stirpium (si fieri potest) peritiam habeat; sin minus plurium saltem quibus frequenter utimur.*     *Galen, Lib. De Antidot.*

# PREFACE

In the catalogues of the *Materia Medica*, the productions of the animal and mineral kingdoms bear a small proportion to those of the vegetable.

Though it must be acknowledged that for some time past the medicinal uses of vegetable simples have been less regarded by physicians than they were formerly, which probably may be ascribed to the successive discoveries and improvements in chemistry;

it would however be difficult to shew that this preference is supported by any conclusive reasoning drawn from a comparative superiority of Chemicals over *Galenicals*, or that the more general use of the former has actually led to a more successful practice.

Although what may be called the herbaceous part of the *Materia Medica*, as now received in the British pharmacopoeias, comprises but a very inconsiderable portion of the vegetable world; yet limited as it now is, few medicinal practitioners have a distinct botanical knowledge of the individual plants of which it is composed, though generally, well acquainted with their effects and pharmaceutical uses.

But the practitioner, who is unable to distinguish those plants which he prescribes, is not only subjected to the impositions of the ignorant and fraudulent, but must feel a dissatisfaction which the inquisitive and philosophic mind will be anxious to remove, and to such it is presumed Medical Botany, by collecting and supplying the information necessary on this subject, will be found an acceptable and useful work; the professed design of which is not only to enable at the reader to distinguish with precision all those plants which are directed for medicinal use by the Colleges of London and Edinburgh, but to furnish him at the same time with a circumstantial detail of their respective virtues, and of the diseases in which they have been most successfully employed by different writers.

A distinctive and characteristic knowledge of natural objects mould certainly precede the consideration of their different properties and qualities; and with respect to plants, this knowledge is seldom to be adequately attained by a mere verbal description: accurate delineations therefore, become necessary, and this department is committed to Mr, Sowerby, an artist of established reputation, whole talents are not less conspicuous in the correctness than in the beauty of his designs.

It is justly a matter of surprise, that notwithstanding the universal adoption of the Linnaean system of Botany, and the great advances made in natural science, the works of Blackwell and Sheldrake should still be the only books in this country in which copper-plate figures of the medicinal plants are professedly given; while splendid foreign publications of them, by Regnault, Zorn, and Plenk, have appeared in the space of a very few years. These works however are far from superseding that now offered to the public; for without resorting to the invidious talk of pointing out their errors and imperfections, the author has the satisfaction of having exhibited Icons of several rare and valuable plants, which have never been completely figured in any preceding work whatever: and by subjoining some account of the botanical and medical history of each species, curiosity is more fully gratified, and a double interest is excited in the mind of the student.

Respecting the uses of Simples, the opinion of Oribasius will not be disputed, viz. "*Simplicium medicamentorum,& facultatum qua in eis insunt, cognitio ita necessaria est, ut sine ea nemo rite medicari queat*" and it is a lamentable truth, that our experimental knowledge of many of the herbaceous simples is extremely defective; for as writers on the *Materia Medica* have usually done little more than copy the accounts given by their predecessors, the virtues now ascribed to several plants are wholly referable to the authority of Dioscorides.

It is however hoped that the medical reader will find what relates to this part of the work as complete as the flow progressive state of experience in physic will admit: with this intention, facts and opinions have been industriously collected from various authorities; and those adduced by Professor Murray, and the works of the late Dr. Cullen, have furnished the largest contribution.

The publication of this work in monthly numbers has afforded the author an opportunity of knowing already the sentiments entertained of it, by several Gentlemen of great medical and botanical authority; from whose unsolicited communications he has derived considerable assistance, and for whose friendly suggestions he desires to make his most grateful acknowledgements.

*(from the preface of first edition of Woodville work.)*

### Note on the plates

While today we may look at these pattern plate books as works of art, it is important to remember their initial function was to provide accurate representations of natural phenomena to promote the study of natural history. These notated hand-colored plates were meant to provide guidance in terms of color and shading to others involved in the printing process.

Details regarding the plates: copper plate engravings in black printing ink, hand colored with watercolor, with iron gall ink and graphite inscriptions.

*Duplex est dos libelli.*

# BIOGRAPHIES

**William Woodville (1752-1805)** was a scientist born into a Quaker family at Cockermouth, Cumberland. He stusies medicine at the university of Edinburgh in Scotland, graduating MD on 12 September 1775. He start his work in European mainland before entering practice in his native country. After he became a famous physician and in 1791 was elected director of the prestigious Smallpox and Inoculation hospitals, St. Pancras in London. With this assignment he appropriated two acres at Battle bridge , near the hospital, where he established a botanical garden, which was maintained at his expenses. In the same year he realize the three volume treatise, *Medical Botany*. In this work , Woodville described, with illustrations and accounts of therapeutic effects all know medicinal plants at this time.

▲ William Woodville M.D. portrait

▼ James Sowerby portrait by Heaphy (1816)

**James Sowerby (1757-1822)** was an English naturalist and illustrator. Trained at the Royal Academy He gained a reputation as a fine illustrator for his contributions to works on botany and other fields of natural science. His reputation was such that he later was involved in several multi-volumed sets published over several years. His family, including both sons and daughters, were involved in his publication endeavors.

# MEDICINAL PLANTS

**M**edicinal plants have been identified and used from prehistoric times. Plants make many chemical compounds for biological functions, including defence against insects, fungi and herbivorous mammals. Over 12,000 active compounds are known to science. These chemicals work on the human body in exactly the same way as pharmaceutical drugs, so herbal medicines can be beneficial and have harmful side effects just like conventional drugs. However, since a single plant may contain many substances, the effects of taking a plant medicine can be complex.

The earliest historical records of herbs are found from the Sumerian civilisation, where hundreds of medicinal plants including opium are listed on clay tablets. The Ebers Papyrus from ancient Egypt describes over 850 plant medicines. Drug research makes use of ethnobotany to search for pharmacologically active substances in nature, and has in this way discovered hundreds of useful compounds. These include the common drugs aspirin, digoxin, quinine, and opium.The compounds found in plants are of many kinds, but most are in four major biochemical classes, the alkaloids, glycosides, polyphenols, and terpenes.

## HISTORY

Plants, including many now used as culinary herbs and spices, have been used as medicines from prehistoric times. Spices have been used partly to counter food spoilage bacteria, especially in hot climates, and especially in meat dishes which spoil more readily. Angiosperms (flowering plants) were the original source of most plant medicines. Human settlements are often surrounded by weeds useful as medicines, such as nettle, dandelion and chickweed. Some animals such as non-human primates, monarch butterflies and sheep ingest medicinal plants to treat illness.

Plant samples from prehistoric burial sites are among the lines of evidence that Paleolithic peoples had knowledge of herbal medicine. For instance, a 60.000-year-old Neanderthal burial site, "Shanidar IV", in northern Iraq has yielded large amounts of pollen from 8 plant species, 7 of which are used now as herbal remedies. The deliberate placement of flowers has been challenged. Paul B. Pettitt has stated that the *"deliberate placement of flowers has now been convincingly eliminated"*, noting that *"A recent examination of the microfauna from the strata into which the grave was cut suggests that the pollen was deposited by the burrowing rodent Meriones persicus, which is common in the Shanidar microfauna and whose burrowing activity can be observed today"*. A mushroom was found in the personal effects of Ötzi the Iceman, whose body was frozen in the Ötztal Alps for more than 5,000 years. The mushroom was probably used to treat whipworm.

## ANCIENT TIMES

In ancient Sumeria, hundreds of medicinal plants including myrrh and opium are listed on clay tablets. The ancient Egyptian Ebers Papyrus lists over 800 plant medicines such as aloe, cannabis, castor bean, garlic, juniper, and mandrake.

From ancient times to the present, Ayurvedic medicine as documented in the Atharva Veda, the Rig Veda and the Sushruta Samhita has used hundreds of pharmacologically active herbs and spices such as turmeric, which contains curcumin. The Chinese pharmacopoeia, the *Shennong Ben Cao Jing* records plant medicines such as chaulmoogra for leprosy, ephedra, and hemp. This was expanded in the Tang Dynasty *Yaoxing Lun*.

In the fourth century BC, Aristotle's pupil Theophrastus wrote the first systematic botany text, *Historia plantarum*. In the first century AD, the Greek physician Pedanius Dioscorides documented over 1000 recipes for medicines using over 600 medicinal plants in *De materia medica*; it remained the authoritative reference on herbalism for over 1500 years, into the seventeenth century.

## MIDDLE AGES

In the Early Middle Ages, Benedictine monasteries preserved medical knowledge in Europe, translating and copying classical texts and maintaining herb gardens. Hildegard of Bingen wrote *Causae et Curae* ("Causes and Cures") on medicine.

In the Islamic Golden Age, scholars translated many classical Greek texts including Dioscorides into Arabic.

Herbalism flourished in Baghdad and in Al-Andalus. Abulcasis (936–1013) of Cordoba wrote *The Book of Simples*, and Ibn al-Baitar (1197–1248) recorded hundreds of medicinal herbs such as *Aconitum*, nux vomica, and tamarind in his *Corpus of Simples*. Avicenna included many plants in his 1025 *The Canon of Medicine*. Abu-Rayhan Biruni, Ibn Zuhr, Peter of Spain, and John of St Amand wrote further pharmacopoeias.

## EARLY MODERN

The early modern period saw the flourishing of illustrated herbals across Europe, starting with the 1526 *Grete Herball*. John Gerard wrote his famous *The Herball or General History of Plants* in 1597, based on Rembert Dodoens, and Nicholas Culpeper published his *The English Physician Enlarged*. Many new plant medicines arrived in Europe as products of Early Modern exploration and the resulting Columbian Exchange. In Mexico, the sixteenth century *Badianus Manuscript* described medicinal plants available in Central America.

# PIANTE OFFICINALI

Una **pianta officinale** è un organismo vegetale usato nelle officine farmaceutiche per la produzione di specialità medicinali. Sono considerate piante officinali piante medicinali, aromatiche e da profumo inserite negli elenchi specifici e nelle farmacopee dei singoli paesi. Il numero e il tipo di piante officinali varia da paese a paese a seconda delle tradizioni. Il più comune utilizzo di piante officinali è quello di correttori del gusto: molti farmaci o preparati farmaceutici hanno originariamente un gusto sgradevole, che quindi viene "corretto" con l'aggiunta di sostanze di origine vegetale. Le piante officinali, ad esempio, sono quelle usate per conferire a sciroppi o a caramelle il gusto di fragola, arancia, limone, ecc.

Nel linguaggio comune spesso si sovrappone l'uso dei termini pianta medicinale con pianta officinale, termini che legalmente indicano due diverse entità; il termine officinale è un termine esclusivamente procedurale e indica quelle piante inserite all'interno di elenchi ufficiali come utilizzabili dalle officine farmaceutiche, a prescindere dal fatto che queste piante abbiano o meno proprietà di tipo medicinale. Il termine pianta medicinale indica invece quelle piante che contengono sostanze utilizzabili direttamente a scopo terapeutico o come precursori in emisintesi che portino a sostanze attive. È quindi chiaro che una pianta può essere officinale in un paese e non in un altro, a seconda delle regolamentazioni, ma essa sarà una pianta medicinale a prescindere dalle leggi. Una **pianta medicinale**, secondo l'Organizzazione Mondiale della Sanità (OMS), è un organismo vegetale che contiene in uno dei suoi organi sostanze che possono essere utilizzate a fini terapeutici o che sono i precursori di emisintesi di specie farmaceutiche.

Si va sempre più affermando il concetto di *fitocomplesso*, quale insieme di sostanze di origine vegetale non riproducibili per sintesi chimica. Il fitocomplesso va inteso come l'insieme di una quantità di principi attivi, noti e non, farmacologicamente attivi, e di sostanze che aiutano l'azione dei primi, pur essendo di per sé queste ultime farmacologicamente inattive. L'insieme delle interazioni dei primi (i principi attivi) e dei secondi (i coadiuvanti) determina le azioni note del fitocomplesso. Sempre secondo l'OMS circa il 25% dei moderni farmaci usati in USA sono di origine vegetale; inoltre sono 7.000 circa i composti medici, presenti nella moderna farmacopea, derivati da piante.

## PIANTA MEDICINALE E OFFICINALE

Nel linguaggio comune si sovrappone l'uso dei termini pianta medicinale con pianta officinale, che indica piante utilizzate nelle officine farmaceutiche per la produzione di specialità medicinali. Questa definizione è però abbastanza riduttiva, e l'utilizzo in ambienti accademici del termine pianta medicinale non fa più riferimento esclusivamente ad un utilizzo a scopo terapeutico delle sostanze contenute nelle piante, bensì dell'utilizzo della pianta o di estratti da essa derivati a scopo terapeutico.

## PIANTE IN PERICOLO

Un'indagine scientifica internazionale promossa dall'OMS all'inizio degli anni novanta, ha rilevato un numero di circa sessantamila specie vegetali, utilizzabili per la cura delle malattie, in forte pericolo di estinzione, di cui trecentosettantaquattro in Italia. Questo fatto richiede una maggiore attenzione alle piante medicinali, non solo quelle utilizzate nelle emisintesi, ma anche quelle che forniscono naturalmente componenti attivi applicabili nell'ambito della fitoterapia.

## CENNI STORICI

Con l'introduzione dell'agricoltura si rese necessaria una maggiore attenzione alla vita delle piante e questo fu il punto di partenza della conoscenza, anche medica, delle caratteristiche delle piante stesse. Il più antico documento medico, per ora rintracciato, è il *"papiro di Ebers"*, risalente al 1500 a.C. Gli egizi facevano largo uso di medicamenti di natura vegetale, in particolar modo conoscevano le proprietà della maggiorana, dell'edera, della mirra.

Nell'antica Grecia, le conoscenze sulle piante si mescolarono con le teorie filosofiche sulle stesse. Uno dei più importanti studiosi fu Eracleide, il quale sperimentò nuove ricette, riprese in seguito da Celso. Le radici studiate e messe in vendita vennero definite *"farmacopoli"* e si basavano soprattutto sulle nozioni contenute nei testi medici scritti da Ippocrate (V secolo a.C.) e in quelli botanici scritti da Teofrasto.

Nell'antica Roma, già nel I secolo d.C. vennero impiantati orti chiamati medicinali, in quanto si coltivavano piante sfruttate per le varie terapie mediche.

Nel IX secolo d.C., in Sicilia, grazie ai Saraceni furono introdotte nuove tecniche idrauliche e di irrigazione che consentirono l'introduzione di nuove piante officinali. Gli arabi diedero un grande impulso sia all'alchimia sia alla chimica, che ebbe ripercussioni nello sviluppo farmaceutico di tinture e distillati. Gli arabi furono i primi ad organizzare una farmacopea, quindi un elenco di ricette descriventi le proporzioni e le composizioni chimiche. Ai secoli XI, XII, XIII, risalgono i primi testi farmaceutici, in cui confluirono le influenze greche, romane e arabe, sintetizzate nella definizione delle operazioni fondamentali: lozione, decozione, infusione e triturazione. In questo periodo si diffuse l'uso delle spezie e delle droghe e la Scuola salernitana introdusse assieme alle pratiche chirurgiche anche un antesignano dell'anestesia, la *spongia sonnifera*, imbevuta di oppio, succo di mandragora e di giusquiamo che doveva essere aspirata dal paziente. La Scuola di Salerno si distinse anche per la grande perizia nel selezionare le erbe, sulle quali abbondano indicazioni terapeutiche che si sono dimostrate efficaci ancora ai nostri tempi, valga per tutte l'insegnamento che diceva: «*Purga l'isopo dalle flemme il petto*», che ha un'azione benefica sulle bronchiti e sulle affezioni respiratorie.

La botanica intesa come scienza nacque solo agli inizi del Cinquecento, grazie alle scoperte geografiche e alla introduzione della stampa. Si diffusero, in questo periodo i primi erbari secchi e nel 1533 a Padova fu istituita la prima cattedra di "botanica sperimentale".

Pietro Andrea Mattioli redasse nel 1554 il più significativi testo di medicina e di botanica dell'epoca. Nel Seicento Pierre Magnol inserì nella classificazione l'intuizione delle famiglie, suddividendo il mondo vegetale in settantasei gruppi.

Nel secolo successivo una grande spinta al progresso della botanica fu effettuata dallo svedese Carl von Linné, che identificò le specie viventi dividendole in basi alle classi, agli ordini e ai generi.

Da allora l'evoluzione è stata continua.

# 3
# THE PLATES
## LE TAVOLE

PLANTS, MEDICINAL

HAND-COLORED ENGRAVING BY

JAMES SOWERBY

PUBLISHED BY

DR. WOODVILLE.

# PLATES LIST OF ILLUSTRATIONS

260. Aloe perfoliata Socotorina - Socotorine Aloe
261. Convallaria Polygonatum - Common Solomon's Seal
262. Iris Florentina - Florentine Orris or Iris
263. Iris Pseudacorus- Yellow water Flag
264. Orchis Mascula - Male Orchis
265. Calamus Rotang - Rotang Cane
266. Saccharum Officinarum - Common Sugar Cane
267. Polypodium Filix mas- Male Polypody or common male Fern
268. Polypodium Vulgare - Common Polypody
269. Asplenium Scolopendrium - Harts-tongue
270. Asplenium Trichomanes – Common Maidenhair or Spleen-wort
271. Lichen Islandicus – Eryngo-leaved Lichen
272. Lichen Caninus – Ash-coloured ground Liverwort
273. Boletus Igniarius - Touchwood Boletus or Argaris
274. Ipecacuanha - Ipecacuan
275. Pinus Balsamica - Balm of Gilead Fir
276. Quercus Infectoria - Staining Oak
277. Salix Caprea - Round leaved Sallow
278. Salix Alba - Common white Willow
279. Solidago Virgaurea - Common golden-rod
280. Cephaelis Ipecacuanha - Ipecacuan
281. Cocculus Palmatus - Plmated Cocculus or Calumba Plant
282. Cinchona Oblongifolia
283. Cinchona Cordifolia
284. Pyrola Umbellata - Umbel-flowered winter-green
285. Scrophularia Nodosa - Knobby-rooted Figwort
286. Pterocarpus Erinaceus - African Pterocarpus or Kino Tree
287. Myloxylon Peruiferum - Sweet-smelling Balsam Tree
288. Diosma Crenata - Crenated Diosma
289. Ranunculus Flammula - Lesser spear-wort Crowfoot
290. Melaleuca Cajuputi - Lesser Cajeput Tree
291. Quassia Excelsa - Lofty or Ash-leaved Quassia
292. Linum Catharticum - Purging Flax or Millmountain
293. Lythrum Salicaria - Loose-strife or purple willow Herb
294. Rhus Toxicodendron - Pubescent poison-oak, Sumach
295. Croton Tiglium - Purging Croton
296. Euphobia Officinarum - Officinal Euphorbium or Spurge
297. Stalagmitis Cambogioides - Camboge Tree
298. Rheum Undulatum - Waved-leaved or Chinese Rhubarb
299. Laurus Cassia -Cassia Tree
300. Humulus Lupulus - Hop
301. Piper Cubeba - Cubebs or Java Pepper
302. Aloe Vulgaris - Yellow-flowered Aloe
303. Triticum Hybernum - Winter or Lammas Wheat
304. Roccella Tinctoria - Dyers Lichen, rock-moss or Orchal
305. Fucus Vesiculosus - Bladder-fucus or Bladder-wrack
306. Roswellia Serrata - Serrated Boswellia or Gum-olibanum Tree
307. Bonplandia Trifoliata - Trhee-leaved Bonplandia
308. Dryobalanops Camphora - Camphor Dryobalanops or Camphor Tree
309. Kramera Triandra - Triandous or Peruvian Kramera

*Sambucus Ebulus*

Published by W. Phillips July 1st 1809.

212- *Sambucus Ebulus - Dwarf Elder*
*Order Dumosae*

*Rhus Coriaria*

213- *Rhus Coriara - Elm-leaved Sumach*
*Order Dumosae*

*Amyris gileadensis.*

Published by W. Phillips July 1st 1814

214- *Amyris Gileadensis - Balsam or Gilead Amyris*
*Order Dumosae*

215.

*Toluifera Balsamum*

Published by W. Phillips. July 1ˢᵗ 1809.

*215- Toluifera Balsamum - Balsam of Tolu Tree*
Order Dumosae

*216* 

*Copaifera officinalis*

Published by V. Phillips August 1804

216- *Copaifera Officinalis - Balsam of Copaiva Tree*
*Order Dumosae*

*Æsculus Hippocastanum*

Published by V. Phillips August 1st 1823

217- *Aesculus Hippocastanum - Common horse Chesnut*
*Order Trihilatae*

*Tropaeolum majus*

Published by W. Phillips, August 1st 1819.

218 - *Tropaeolum Majus - Greater Indian Cress or Nasturtium*
Order *Trihilatae*

*Berberis vulgaris*

*Published by W. Phillips, August 1, 1819.*

219- *Berberis Vulgaris - Common Barberry*
*Order Trihilatae*

220- *Swietenia Mahagoni* - *Mahogany Tree*
Order *Trihilatae*

*Ricinus communis*
Published by W. Phillips 1809

*Croton Cascarilla*
Published by W. Phillips 1809

*Clutia Elaterina & Cascarilla*

*Siphonia Sanlana*
Published by W. Phillips 1809

*221-222-223-224- Tricoccae Various*
Order Tricoccae

225- *Thea - Tea Tree*
*Order Tricoccae*

*Wintera aromatica*

Published by W. Phillips, Oct.r 1.st 1809.

*226- Wintera Aromatica - Winter's Bark Tree*
*Order Tricoccae*

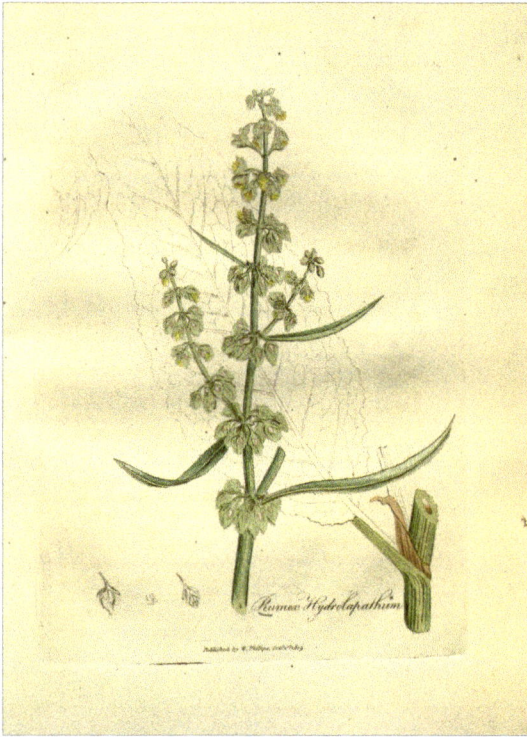

*Salsola Kali*

*Chenopodium Vulvaria*

*Rumex Hydrolapathum*

*Rumex Acetosa*

227-228-229-230- *Oleraceae Various*

*Order Oleraceae*

*231- Rheum Palmatum - Officinal Rhubarb*

*Order Oleraceae*

*232- Polygonum Bistorta - Greater Bistort or Snakeweed*
*Order Oleraceae*

*Laurus Cinnamomum.*

Published by W. Phillips, Nov.r 1.st 1809.

*233- Laurus Cinnamomum - Cinnamon Tree*
Order Oleraceae

*Laurus Sassafras*

Published by W. Phillips, March 1809.

234 - *Laurus Sassafras* - *Sassafras Tree*
Order *Oleraceae*

235- *Laurus Nobilis- Common Sweet-bay*

Order Oleraceae

*Laurus Camphora*

Published by W. Phillips. Dec.' 1809.

*236- Laurus Camphora - Camphor Tree*
*Order Oleraceae*

*Canella alba.*

Published by W. Phillips, Dec.r 1st 1819.

*237- Canella Alba - Laurel-leaved Canella*
Order Oleraceae

238- *Myristica Moschata - Nutmeg Tree*
Order Oleraceae

*Parietaria officinalis*

*Dorstenia Contrajerva*

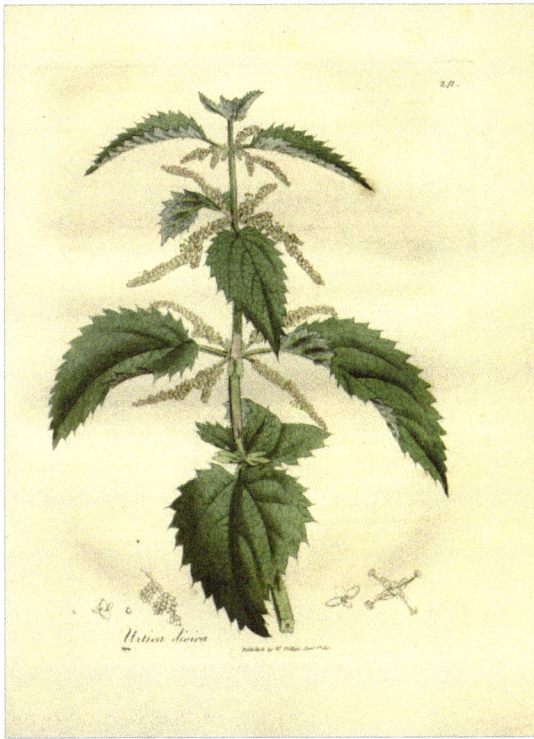

*Urtica dioica*

*Ulmus campestris*

*239-240-241-242- Scabridae Various*

*Order Scabridae*

*Morus nigra*

Published by W. Phillips Jan'y. 'st. alte.

*243- Morus nigra - Common Mulberry Tree*
Order Scabridae

*Ficus Carica*

Published by W. Phillips, Jan.ʳ 1ˢᵗ 1820.

*244- Ficus Carica - Common Fig Tree*
*Order Scabridae*

*Daphne Mezereum*

Published by W. Phillips. Invl's Sabo.

245- *Daphne Mezereum* - *Mesereon*
Order *Vepreculae*

246.

*Piper nigrum.*

Published by W. Phillips, Feb. 1, 1811.

246- *Piper nigrum - Black Pepper*
Order *Piperitae*

*Piper longum*

Published by W. Phillips, Feby 1st 1821.

*247- Piper Longum - Long Pepper*
Order Piperitae

248- *Acarus Calamus - Sweet Flag or Acorus*
Order *Piperitae*

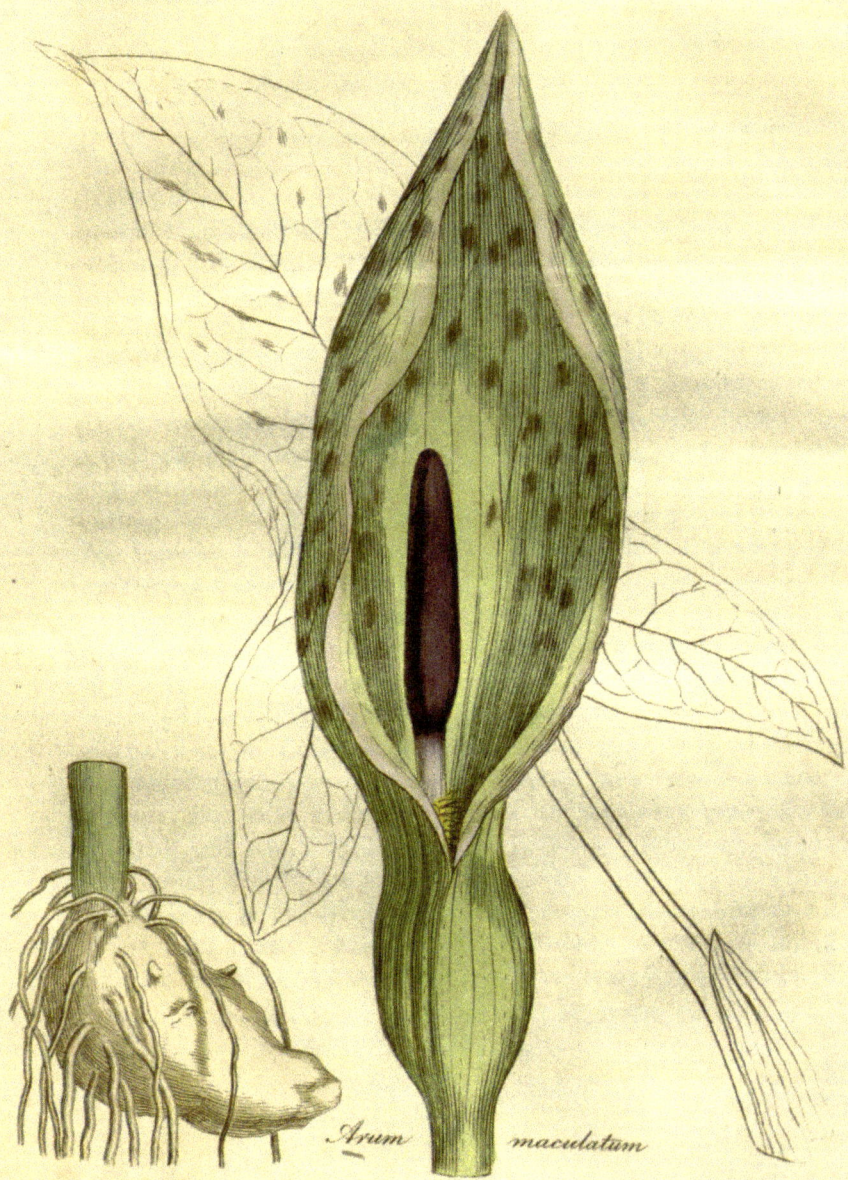

*Arum maculatum*

Published by W. Phillips, ...

**249- Arum Maculatum - Common Arum or Wake-robin**
*Order Piperitae*

250- *Amomum Zingiber - Narrow-leaved Ginger*
*Order Scitamineae*

*Amomum repens   seu Cardamomum.*

Published by W. Phillips, March 1st 1820.

251- *Amomum Repens seu Cardamomum - Officinal Cardamom*
Order Scitamineae

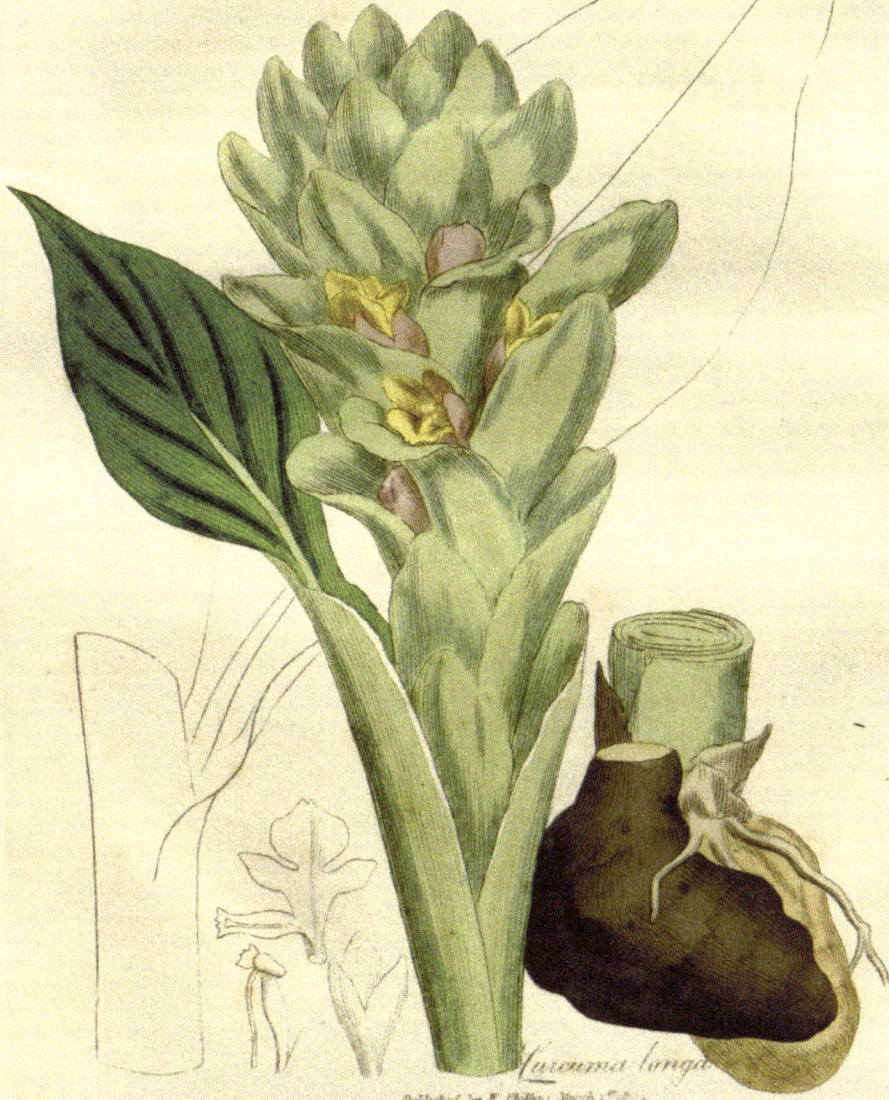

*252- Curcuma Longa - Long-rooted Turmeric*
Order Scitamineae

253.

*Kæmpferia rotunda*

Published by W. Phillips March 1 1810.

253- *Kaempferia Rotunda - Zedoary*
Order Scitamineae

*Lilium candidum*

Published by W. Phillips March 1 1812.

*254- Lilium Candidum - Common white Lily*

*Order Liliaceae*

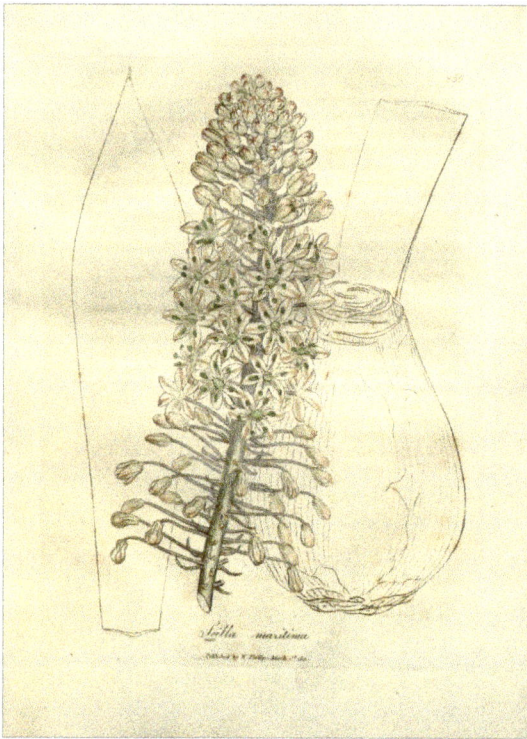

*Scilla maritima.*
Published by W. Phillips March 1st 1812.

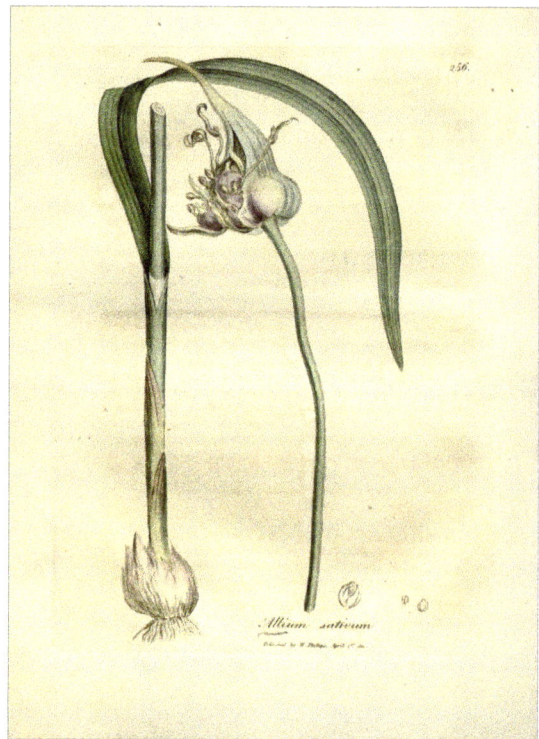

256.

*Allium sativum.*
Published by W. Phillips April 1st 1812.

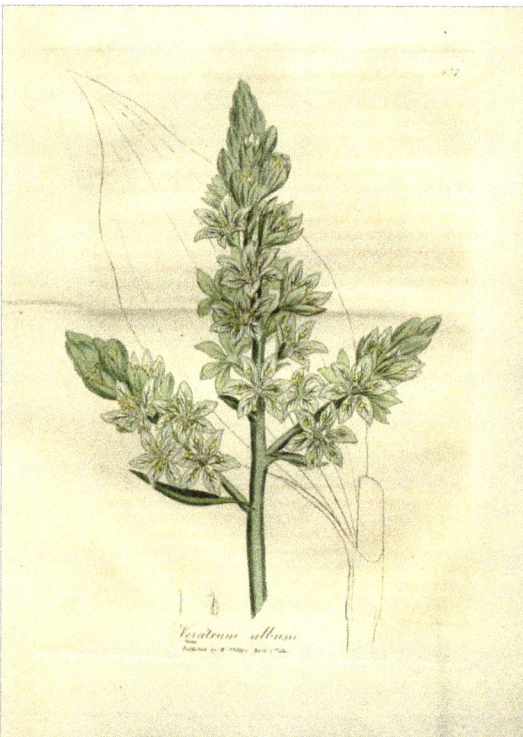

*Veratrum album.*
Published by W. Phillips April 1st 1812.

358.

*Colchicum autumnale.*
Published by W. Phillips April 1st 1812.

*255-256-257-258- Liliaceae Various*
Order Liliaceae

*Crocus sativus*

Published by W. Phillips. April 1. 1810.

259- *Crocus Sativus - Saffron Crocus*
*Order Liliaceae*

*Aloe perfoliata Socotorina*

Published by T. Phillips, April 1st 1826

260- *Aloe perfoliata Socotorina - Socotorine Aloe*
*Order Liliaceae*

*Convallaria Polygonatum.*

Published by W. Phillips, Maxenden

261- *Convallaria Polygonatum - Common Solomon's Seal*
*Order Liliaceae*

*Iris florentina*

Publised by W. Phillips, May 1812

262- *Iris Florentina - Florentine Orris or Iris*
Order Ensatae

*Iris Pseudacorus.*

Published by W. Phillips, May 1812

*263- Iris Pseudacorus- Yellow water Flag*
Order Ensatae

*264- Orchis Mascula - Male Orchis*

*Order Orchideae*

*Calamus Rotang*

Published by W. Phillips, 1 May 1812

265- *Calamus Rotang - Rotang Cane*
*Order Tripedaloideae*

*Saccharum officinarum*

*266- Saccharum Officinarum - Common Sugar Cane*
*Order Gramina*

*Polypodium Filix mas*

Publish'd by W. Phillips, June 1811

267- *Polypodium Filix mas- Male Polypody or common male Fern*
Order Filices

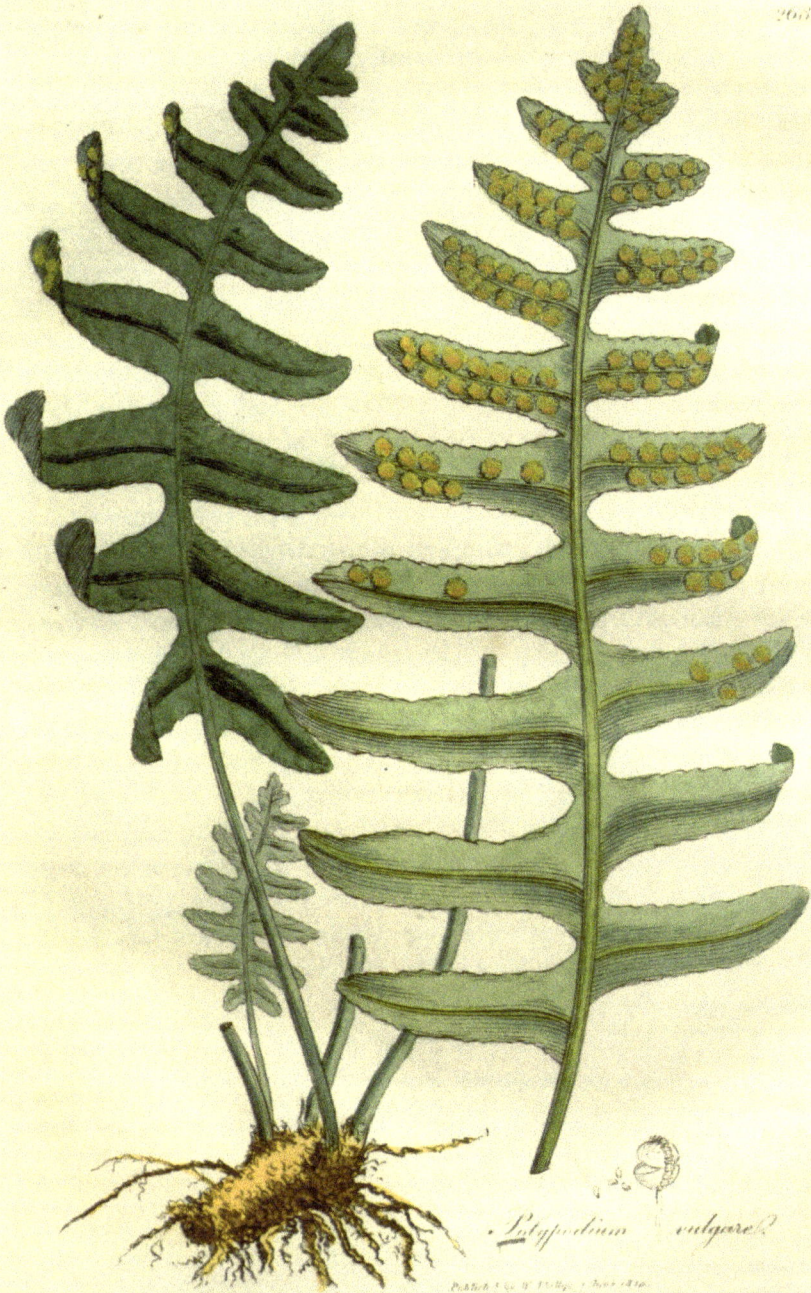

*268- Polypodium Vulgare - Common Polypody*
Order Filices

*269- Asplenium Scolopendrium - Harts-tongue*
Order Filices

270

*Asplenium Trichomanes*

271

*Lichen islandicus*

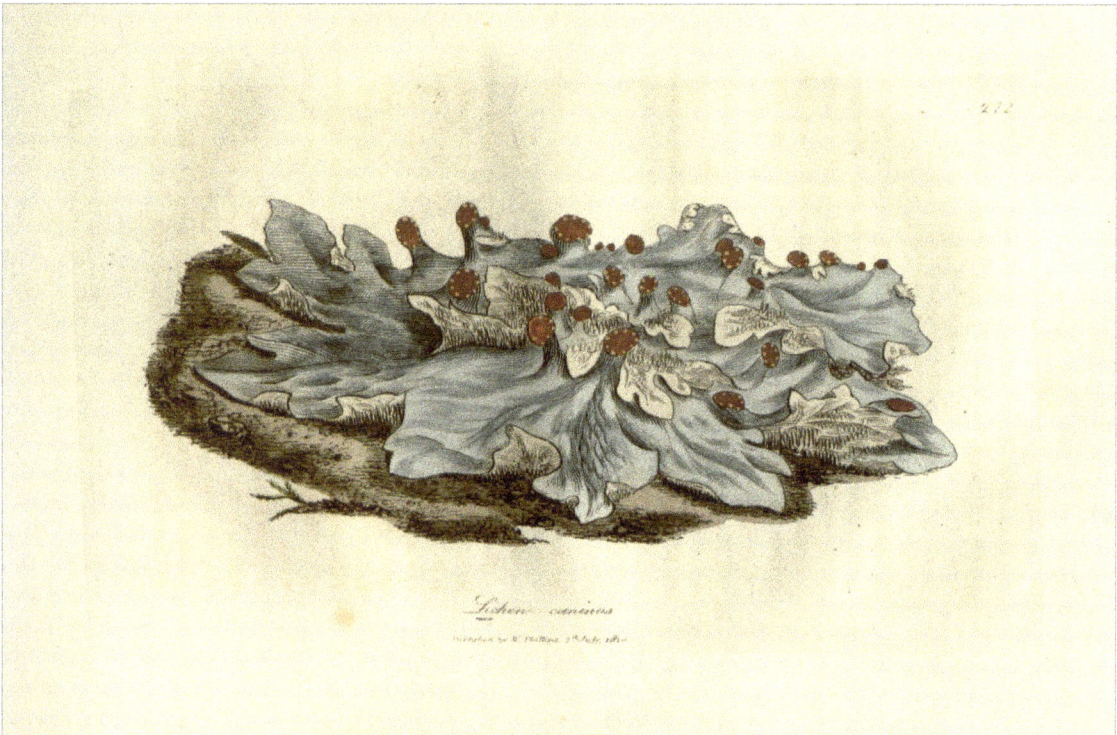

272

*Lichen caninus*

270-271-272- *Filices and Algae various*
Order *Filices and Algae*

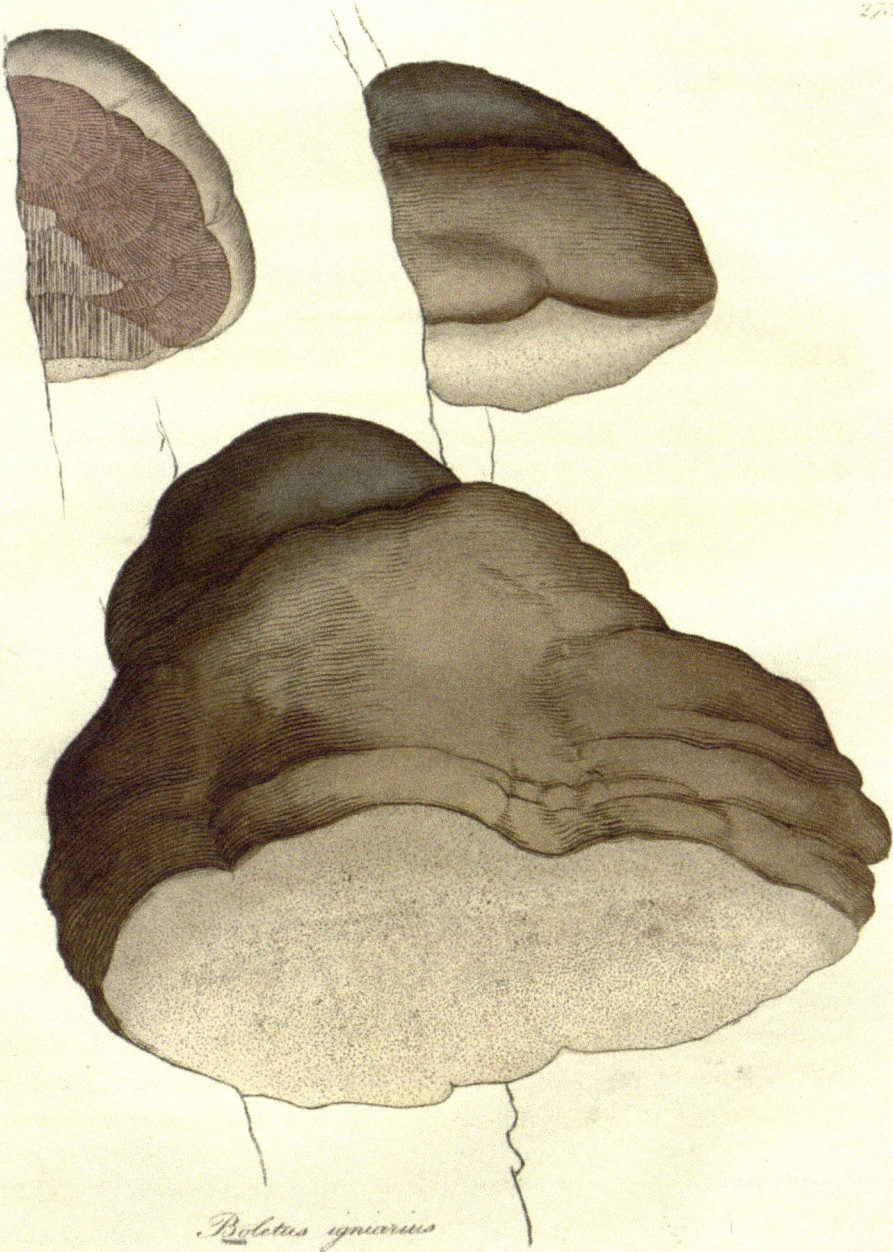

*Boletus igniarius*

Published by W. Phillips, 2d July, 1810

273- *Boletus Igniarius - Touchwood Boletus or Argaris*
Order Fungi

*Ipecacuanha*

Published by W. Phillips, 1.st July 1810

*Cephaelis Ipecacuanha*

274- *Ipecacuanha - Ipecacuan*
Order Fungi

*Pinus balsamea.*

G. Spratt del.

275- *Pinus Balsamica - Balm of Gilead Fir*
*Order Coniferae*

*Quercus Infectoria.*

G. Spratt del.

*276- Quercus Infectoria - Staining Oak*
Order Amentaceae

Fig. 1.                          Fig. 2.

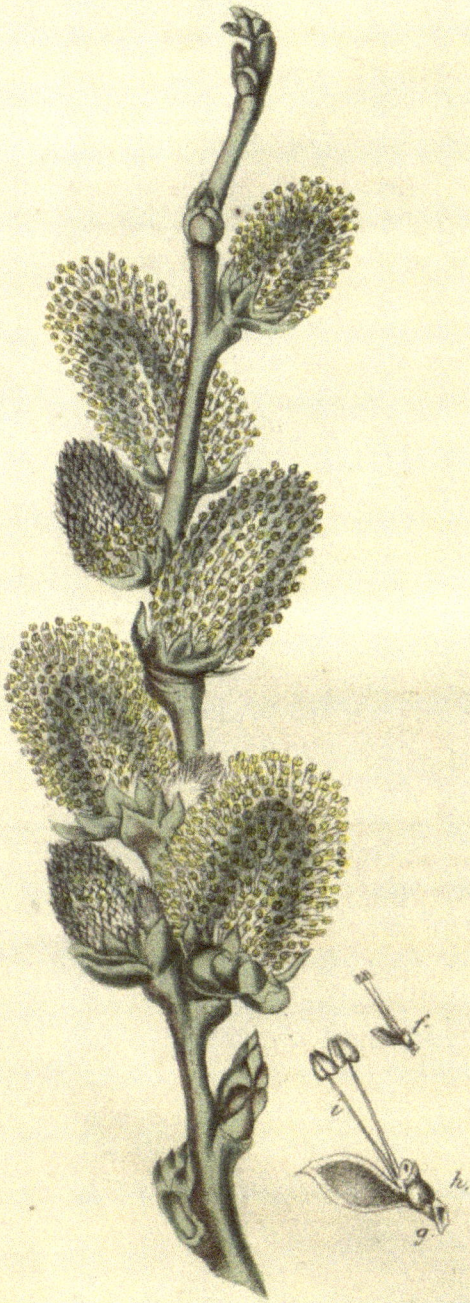

*Salix caprea.*

G. Spratt del.

*277- Salix Caprea - Round leaved Sallow*
*Order Amentaceae*

*Salix alba.*

*278- Salix Alba - Common white Willow*

Order Amentaceae

*Solidago virgaurea.*

G. Sowerby del.

279- *Solidago Virgaurea - Common golden-rod*
*Order Compositae Discoideae*

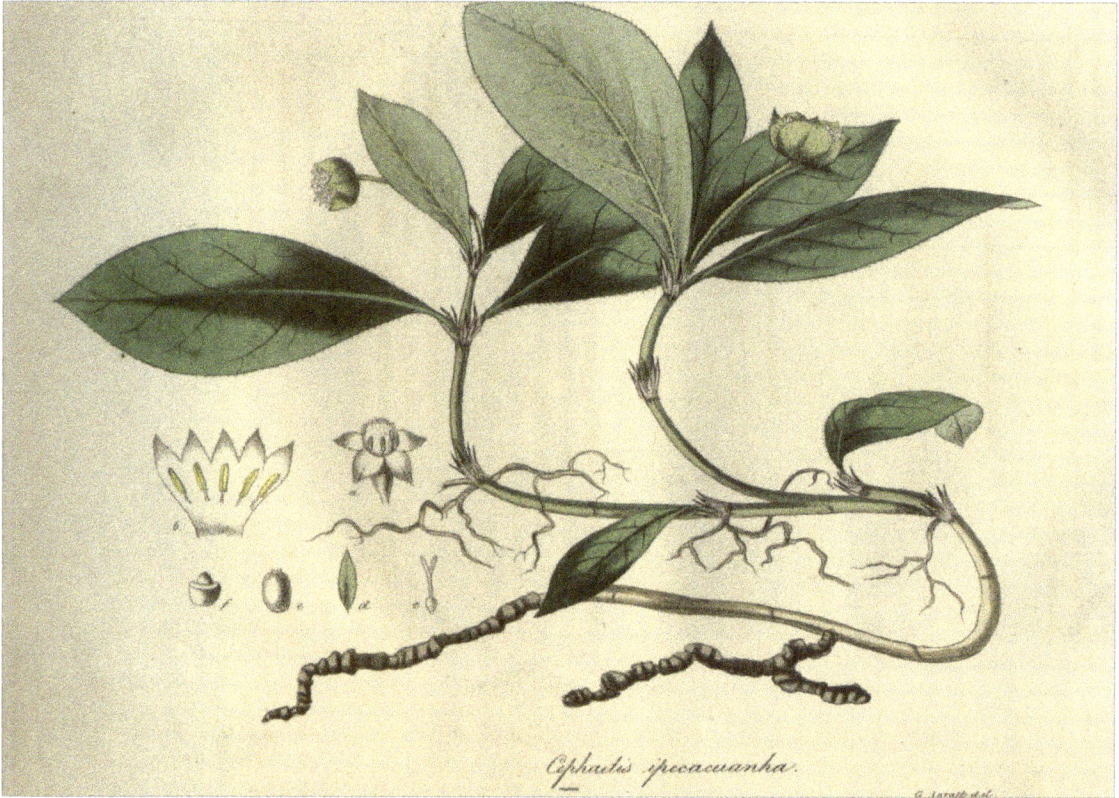

*Cephaelis ipecacuanha.*

G. Soralb del.

**280- Cephaelis Ipecacuanha - Ipecacuan**
*Order Aggregatae*

*Cocculus palmatus.*

281- *Cocculus Palmatus - Plmated Cocculus or Calumba Plant*

*Order Sarmentaceae*

*Cinchona Oblongifolia.*

282- *Cinchona Oblongifolia*
Order *Rubiaceae*

*Cinchona cordifolia.*

G. Spratt del.

283- *Cinchona Cordifolia*
*Order Rubiaceae*

*Pyrola umbellata.*

G.Spratt del.

284- *Pyrola Umbellata - Umbel-flowered winter-green*
*Order Bicornes*

*Scrophularia Nodosa.*

G. Spratt del.

285- *Scrophularia Nodosa - Knobby-rooted Figwort*
*Order Personatae*

*Pterocarpus erinaceus.*

C. Spratt del.

286- *Pterocarpus Erinaceus - African Pterocarpus or Kino Tree*
Order Papilionaceae

*Myroxylon Peruiferum.*

E. Spratt del.

287- *Myloxylon Peruiferum - Sweet-smelling Balsam Tree*
Order *Lomentaceae*

*Diosma Crenata.*

G. Spratt, del.

288- *Diosma Crenata - Crenated Diosma*
Order Multisiliquae

*Ranunculus flammula*

**289- *Ranunculus Flammula* - *Lesser spear-wort Crowfoot***
*Order Multisiliquae*

*Melaleuca cajuputi.*

G. Spratt del.

290- *Melaleuca Cajuputi - Lesser Cajeput Tree*
*Order Hesperideae*

*Quassia excelsa.*

a Spratt del

291- *Quassia Excelsa - Lofty or Ash-leaved Quassia*
*Order Gruinales*

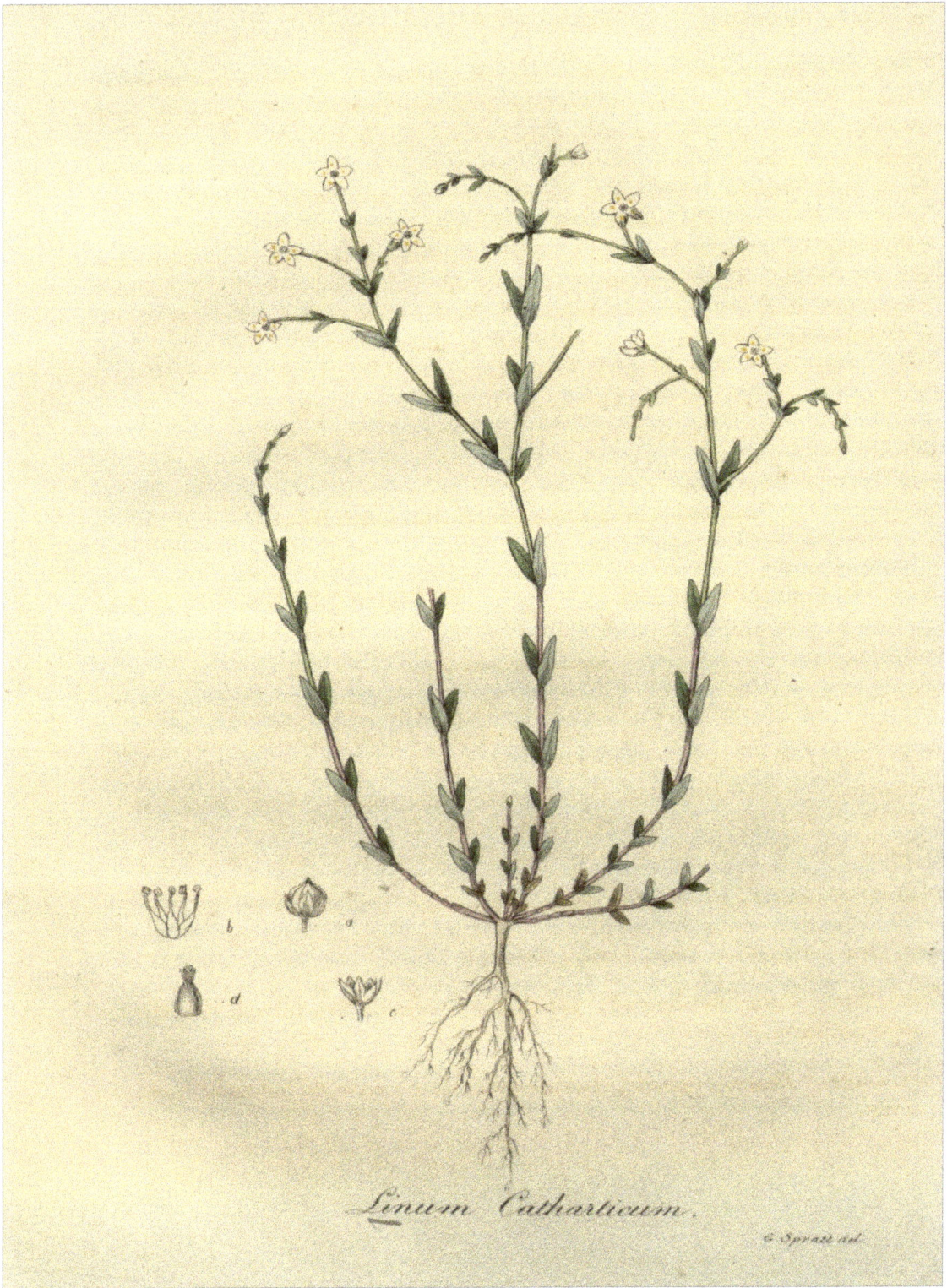

*Linum Catharticum.*

G. Spratt del

292- *Linum Catharticum - Purging Flax or Millmountain*
*Order Gruinales*

*Lythrum salicaria.*

G. Spratt dd.

293- *Lythrum Salicaria - Loose-strife or purple willow Herb*
*Order Calycanthemae*

*Rhus toxicodendron.*

*294- Rhus Toxicodendron - Pubescent poison-oak, Sumach*
Order Dumosae

*Croton Tiglium*

*295- Croton Tiglium - Purging Croton*
Order Tricoccae

*Euphorbia officinarum.*

G. Spratt del.

296- Euphobia Officinarum - Official Euphorbium or Spurge
Order Tricoccae

*Stalagmitis cambogioides.*

G.Spratt del.

*297- Stalagmitis Cambogioides - Camboge Tree*
Order Tricoccae

*Rheum undulatum*

C. Spratt del.

298 - *Rheum Undulatum - Waved-leaved or Chinese Rhubarb*
Order Holeraceae

*Laurus cassia.*

G. Spratt del.

299- *Laurus Cassia -Cassia Tree*
*Order Holeraceae*

*Humulus lupulus.*

C. Spratt del.

300- *Humulus Lupulus - Hop*
Order Scabridae

*Piper cubeba.*

G. Spratt. del.

301- *Piper Cubeba - Cubebs or Java Pepper*
*Order Piperitae*

*Aloe Vulgaris.*

*302- Aloe Vulgaris - Yellow-flowered Aloe*
*Order Liliaceae*

*Triticum Hybernum.*

*Avena Sativa.*       *Hordeum Distichon.*

303- *Triticum Hybernum - Winter or Lammas Wheat*
Order Gramina

*Roccella tinctoria.*

C. Spratt del.

304 - *Roccella Tinctoria* - *Dyers Lichen, rock-moss or Orchal*
*Order Algae*

*Fucus Vesiculosus.*

G. Spratt del.

305- *Fucus Vesiculosus - Bladder-fucus or Bladder-wrack*
*Order Algae*

*Boswellia serrata.*

G. Spratt del.

306- *Roswellia Serrata - Serrated Boswellia or Gum-olibanum Tree*
*Order Therebintaceae*

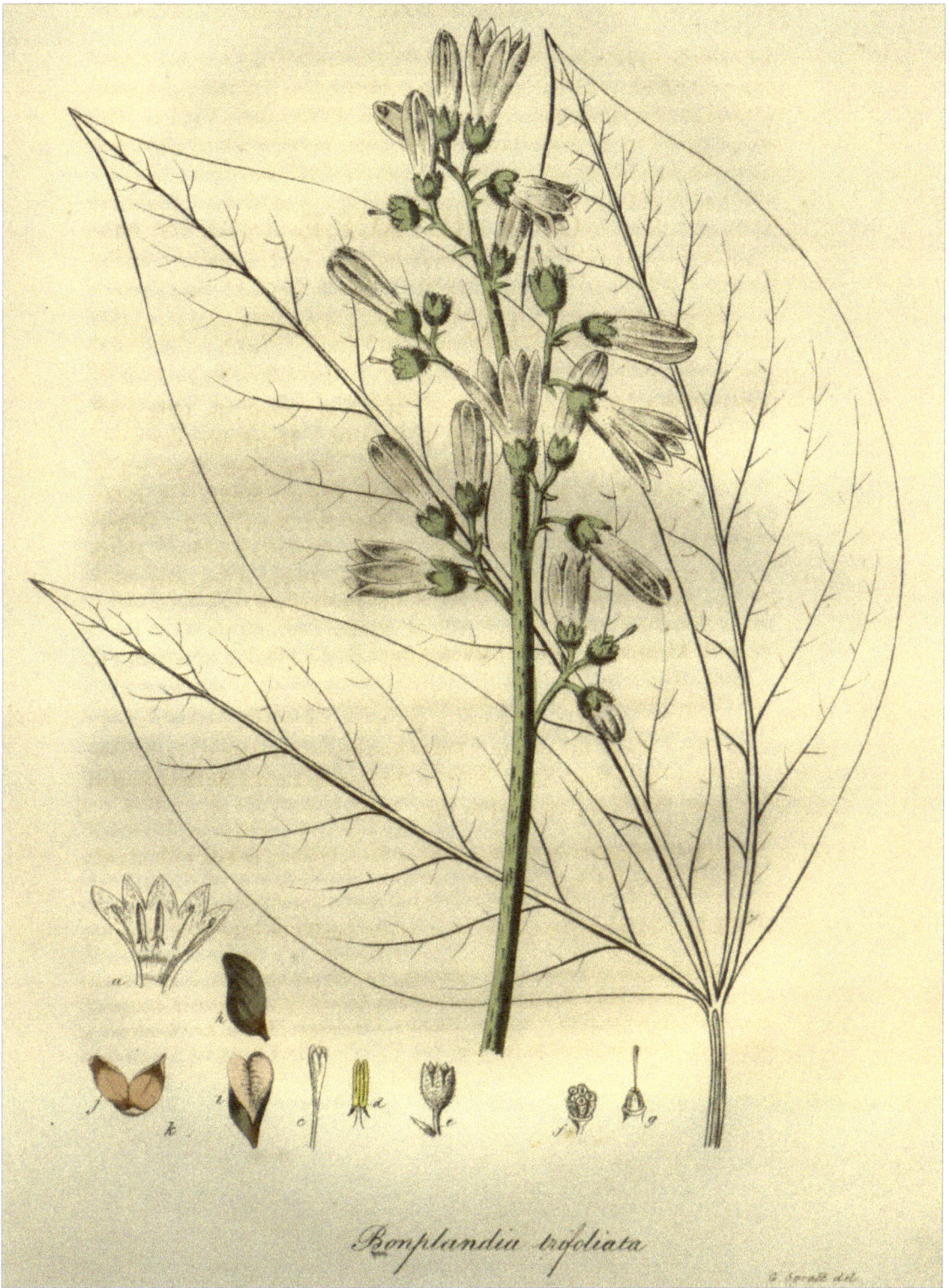

*Bonplandia trifoliata*

G. Sorelli del.

*307- Bonplandia Trifoliata - Trhee-leaved Bonplandia*
*Order Rutaceae*

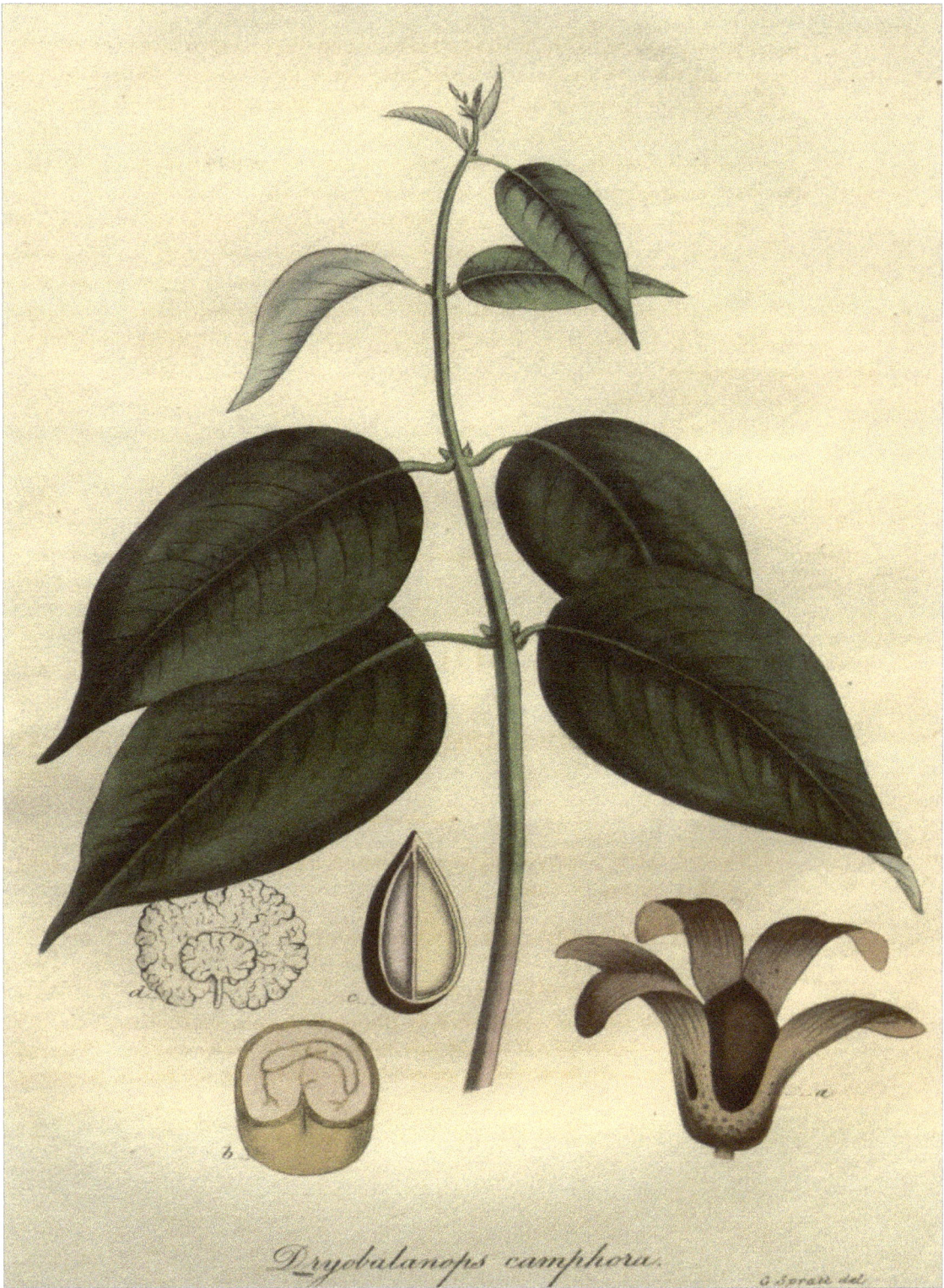

*Dryobalanops camphora.*

G. Sprott del.

**308 - *Dryobalanops Camphora* - *Camphor Dryobalanops or Camphor Tree***
*Order Dipterocarpeae*

*Krameria Triandra.*

G. Sprett del.

309- *Kramera Triandra - Triandous or Peruvian Kramera*
*Order Polygaleaea*

## DARWIN'S VIEW SERIES

Actually, the world from Darwin's point of view. The new series of specifically dedicated to the animal, vegetable and mineral world. A great review of nature through his most beautiful and fascinating images, taken from ancient tomes and essays about nature, made by the greatest individuals, artists and scientists together. Not only that, "Darwin's view" will involve yourself through the description of the stories, with facts and images of the exotic and romantic travels, made by the great explorers and brilliant scientists of the past, starting with the epic one on the HMS Beagle of our beloved and legendary Charles Robert Darwin!

# CONTENTS